INTERNATIONAL CENTRE FOR MECHANICAL SCIENCES

COURSES AND LECTURES - No. 144

HAYRETTIN KARDESTUNCER
UNIVERSITY OF STORRS, CONNECTICUT

FINITE ELEMENTS METHODS VIA TENSORS

COURSE HELD AT THE DEPARTMENT
OF -MECHANICS OF SOLIDS
JUNE 1972

UDINE 1972

SPRINGER-VERLAG WIEN GMBH

This work is subject to copyright.

All rights are reserved,

whether the whole or part of the material is concerned
specifically those of translation, reprinting, re-use of illustrations,
broadcasting, reproduction by photocopying machine
or similar means, and storage in data banks.

© 1972 by Springer-Verlag Wien

Originally published by Springer-Verlag Wien New York in 1972

ISBN 978-3-211-81224-2   ISBN 978-3-7091-4323-0 (eBook)
DOI 10.1007/978-3-7091-4323-0

Dedicated to (*)

Professor Luigi Sobrero

Father of CISM

(*) The most modest dedication of all.

PREFACE

The text introduces holor algebra with tensor notations to the analysis of discrete elastic systems. Although problems of this nature are often formulated in matrix notations, the elimination of unnecessary difficulties in such notations is worth looking into. Considering that most entities encountered in discrete mechanics observe well-defined rules under the coordinate transformations, their treatment as arithmetic vectors and matrices as done in contemporary methods may not be justified. Consequently, we introduce tensorial entities into discrete mechanics similar to those in continuum mechanics. For instance, the difference between the "stiffness" and "tensor of moduli" is that the first one is determined not only by the physical properties of the material, as is the case in the latter one, but by the geometry of the system as well.

The author is greatly indebted to the Rectors and to the Secretary General of the International Centre for Mechanical Sciences for providing him such an excellent opportunity to present these lectures to such a selected audience in the very academic atmosphere of CISM.

Udine, June 1972.

INTRODUCTION

Not long ago, almost toward the middle of this century the formulation of problems in mechanics (in other sciences as well) began using matrices extensively. Since the entities such as vectors and matrices having large number of parameters are identified by single symbols, formulation of problems became short and coincise. Because of the availability of powerfull computers scientists were encouraged to present a single parameter or a function in analytical expressions by collection of many functions or parameters with certain approximations. This, of course, shifted the formulation of the problems from "exact" to "approximate". Since the definitions of problems are often not exact to begin with, their formulation as well as their solutions being approximate (depending upon the degree of approximation of course) received no objections. To the contrary, it was assumed that an "approximate solution" is better than "no solution". This way of thinking encouraged researchers to attempt to solve many unsolvable problems of yesterday.

Although matrix algebra has a number of merits, it has however its own shortcomings because of many rigid rules and regulations which must be complied with. The conformability of the entities, the order of operations, non-physical characteristics of the equations introduce certain difficulties and the loss of the physical concept of the problem. Since the finite element method in solid mechanics seldom involves the "calculus of matrices", the formulation of the problems relies heavily on rigid rules of matrix algebra which hardly differenciate between the physical and non-physical problems. When an entity does not contain apple and pear or a number of cars in a lot, it would perhaps deserve a different kind of treatment than otherwise.

Granting that Einstein once said, "The physical problems can best be formulated in tensor notations", an attempt should be made to explore such wisdom. Most entities in the finite element method, with the exception of those that establish liaison between them, are invariant under the coordinate transformations[2]. Their expression in one system is often simpler than in an other system, and the equations that contain these entities possess easier solution in certains coordinates. For instance, a matrix might be "full" in one coordinate system yet "loose", "banded" even "diagonal" in the other system. As a matter of fact the solution of a set of simultaneous equations, triangularization, diagonalization of matrices merely involves coordinate transformations.

Consider for instance an equation in the form of

(1) $$Ax = y$$

expressed in one coordinate system.

Let the square matrix A in this equation be non-singular. Then, the solution of this equation implies diagonalization of matrix A by expressing this equation in another coordinate system. Such a coordinate system is often referred to as the "principal coordinates" of the phenomenon

$$A'x' = y' \therefore A' = RAR^*, \ x' = R^{*^{-1}} x, \ y' = Ry$$

where R is the "transformation matrix" between the old and the new coordinate systems.

Among many methods for the solution of Eq. (1) Gauss' and Cholesky's are the most frequently employed in practice. While the first one operates on rows and columns of A the second separates it into upper and lower triangular parts. Such operations are always possible as long as matrix A is real, non-singular and square. Whether it is loose or dense, banded or tridiagonal, whether it possesses some other properties makes no difference.

In the formulation of the problems in continuum by the finite element method, however, regardless of the procedure, the final equation resembles Eq. (1) above. Such an equation is often written as

$$P = K V \qquad (2)$$

in which **P** and **V** are referred to as the generalized force and displacement vectors and the square matrix **K** is the stiffness matrix of the system.

Once such an equation is formed, a common practice today is to refer to Gauss or Cholesky to obtain the unknown displacements. There are many other methods[12] suitable for the solution of this equation which pay attention to certain properties of K such as loose , banded , tridiagonal , well- or ill- conditioned etc... None of them, however, pay attention to the physical properties of K. The reason for it, of course, is the fact that in matrix algebra there is no difference between Eq. (1) and (2). Both are linear simultaneous equations with constant coefficients. Here, where the engineers who have accomplished a great deal in obtaining such a simple equation stop and refer it to the ordinary methods for the solution. One should of course expect that during the formulation of a physical phenomenon in matrix notation, the problem resembles nothing more than a set of simultaneous equations. In order not to lose contact with the physical phenomenon, here, we shall try to formulate the problem in "tensor" notations. In order to establish a smooth transition between the classical matrix method and the tensorial formulation of the finite element method let us introduce the former very briefly.

MATRIX FORMULATION OF THE PROBLEM

Assume that a body is subject to quasi-linear application of concentrated loads. For the sake of simplicity further assume that the surface or body forces are either omitted or transformed into "equivalent" concentrated loads. If σ and ϵ are the state of stress and strain of the body, respectively, the total strain energy observed by the body at the end of the loading sequence would be

(3) $$U = \tfrac{1}{2} \int_\Omega \epsilon^* \, \sigma \, d\Omega$$

Considering that

(3a) $$\sigma = E \, \epsilon$$

represents the stress-strain relationship, and

(4) $$\epsilon = b \, v \therefore \epsilon_{xx} = \frac{\partial v}{\partial x}, \text{ etc}$$

is the strain-displacement relationship, Eq. (3) becomes

(5) $$U = \tfrac{1}{2} \int_\Omega v^* \, \dot{b} \, E \, b \, v \, d\Omega$$

According to Castigliano's theorem[1] (part I), however

(6) $$\frac{\partial U}{\partial v} = p = \int_\Omega b^* \, E \, b \, d\Omega \, v$$

or

(7) $$p = k \, v$$

where

$$k = \int_\Omega b^* E b \, d\Omega \qquad (8)$$

If Ω represents a domain with finite size, then Eq. (7) represents force-displacement relationship in that domain. Eq. (8), then, is referred to as the stiffness matrix of the element.

As it will be shown later on, if a continuous body is divided into n number of finite size of elements, and Eq. (7) is accomplished for each element, then through "displacement compatibilities" at the "nodal" points of the elements the overall matrix equation of the entire body (similar to that of Eq. (2)) can be formed.

Although the formulation of the problem is completed at this stage, evidently contact with the physical problem is lost. Instead, we shall start to reformulate the problem in tensor notations and try not to loose contact with physics[4].

TENSORIAL FORMULATION OF THE PROBLEM

Tensors have been largely utilized in the analysis of electric circuits. Especially, Gabriel Kron [20] has been the very first author to deal with tensors extensively in circuit analysis. He also introduced tensors with his tearing method of analysis (Diakoptics) into discrete systems. Sokolnikoff [5] and many others [21, 22, 23, 24] indicated the power of tensors in continuum mechanics. J.W. Soule [25] employed tensors in piping flexibility analysis. S.F. Borg [26] slightly touched tensors in structural analysis.

When the reader refers to these references, he may notice that tensors are interpreted in different ways in every one of them. Borg avoided the use of indices and dealt only with bivalent tensors. Kron practically referred to every holor as tensor and Fung employed only cartesian coordinates where the distinction between covariance and controvariance properties of tensors disappears. Here we are making a very modest attempt to use tensors with higher valences in the analysis of discrete elastic systems.

Regardless of the method employed for the solution of the problem in continuum mechanics, the results are expected to yield either one of the two tensorial fields : namely, stress and strain. The relationship between these two fields is the well-known Cauchy's equality[5]

(9) $$\sigma^{iq} = E^{ijqr} \epsilon_{jr} \quad i,j,q,r = 1,2,3$$

in which E^{ijqr} is known as the "tensor of moduli" or the generalized Hooke's tensor. Such an equality which is written in three-dimensional physical space can take the following form

(9a) $$\sigma_{iq} = E_{ijqr} \epsilon_{jr}$$

in Cartesian coordinates where the distinction between the "covariant and contravariant" components of tensors disappear[6].

Now, considering that force can be interpreted as the limiting value of stress with area approaching to zero and having the covariant components of displacements, the strain energy of the body can be approximated as

(10) $$U = \tfrac{1}{2} p^i v_i$$

where the summation is done over the discrete points (nodal points) of the body.

Tensorial Formulation of the Problem

Again, according to Castigliano's first theorem

$$\frac{\partial U}{\partial v_i} = p^i = \tfrac{1}{2}\frac{\partial(p^j v_j)}{\partial v_i} = \tfrac{1}{2}\frac{\partial p^j}{\partial v_i}v_j + \tfrac{1}{2} p^j \frac{\partial v_j}{\partial v_i}$$

Since, however,

$$\frac{\partial v_j}{\partial v_i} = 0, 1 \text{ for } i = j\,;\, i \neq j \text{ respectively}$$

therefore

(10a)
$$p^i = \frac{\partial p^j}{\partial v_i} v_j$$

This equation which is written in one coordinate system is also valid in any other coordinate system

$$p^i = k^{ij} v_j \quad \therefore \quad p^{i'} = k^{i'j'} v_{j'}.$$

since

$$p^{i'} = \frac{\partial x_{i'}}{\partial x_i} p_i \quad \therefore \quad v_{j'} = \frac{\partial x_j}{\partial x_{j'}} v_j$$

Thus

$$p^i = \frac{\partial x_i}{\partial x_{i'}} \frac{\partial x_j}{\partial x_{j'}} k^{i'j'} v_j \quad \therefore \quad k^{ij} = \frac{\partial x_i}{\partial x_{i'}} \frac{\partial x_j}{\partial x_{j'}} k^{i'j'}$$

or

$$k^{i'j'} = \frac{\partial x_{i'}}{\partial x_i} \frac{\partial x_{j'}}{\partial x_j} k^{ij}$$

indicates that the "stiffness tensor" assumes contravariant tensor transformation. This transformation is equivalent to the following well-known and largely employed transformation[7]

$$k' = R\, k\, R^*$$

in matrix algebra. Note, however, that while the former transformation can be expressed as either one of the following forms

$$k^{i'j'} = \frac{\partial x_{i'}}{\partial x_i} k^{ij} \frac{\partial x_{j'}}{\partial x_j} = \frac{\partial x_{j'}}{\partial x_j} \frac{\partial x_{i'}}{\partial x_i} k^{ij}, \text{ etc.}$$

the latter assumes "no" other version. Since the coordinate transformations are very largely employed especially between "local" and "global" coordinate systems[8], the advantage of lucidity and flexibility of tensor transformations ought to be acknowledged during the formulation of the problem.

If, now, force and displacement vectors are expressed in terms of their components in the chosen coordinate system, Eq. (10a) takes the following form

(11) $$p^{iq} = k^{ijqr} v_{jr} \qquad q, r = 1, \ldots, N$$

where N represents "the degree of freedom" of the physical domain in which the problem is defined.

It would, perhaps, be worth noticing the similarities between Eqs. (9) and (11). While the former is expressed in three-dimensional physical space, the latter is in n-dimensional mathematical space. Furthermore the first one is continuous in the entire domain and valid in every infinitesimal element, the second is valid at certain discrete points only. While the tensor of Moduli is solemnly dependent upon the physical properties of the material, the "stiffness tensor" is controled by geometry as well. It can also be shown that Eq. (11), similar to that of Eq. (9), possesses "symmetry". For instance, expressing the strain energy of an element as

$$U = \tfrac{1}{2} p^{iq} v_{iq}$$

then

$$k^{ijqr} = \frac{\partial^2 U}{\partial v_{iq} \partial v_{jr}} = \frac{\partial}{\partial v_{iq}} \left(\frac{\partial U}{\partial v_{jr}}\right)$$

Tensorial Formulation of the Problem

Since, however, the order of differentiation bears no importance

$$k^{ijqr} = \frac{\partial}{\partial v_{jr}}\left(\frac{\partial U}{\partial v_{iq}}\right) = k^{jriq}$$

which proves the symmetry.

At this moment, let us go back to Eq. (10) and reformulate the problem by using the second theorem of Castigliano instead of the first theorem. In doing this, the force-displacement relationship will be obtained in reverse order.

Let

$$U = \frac{1}{2} p^{jr} v_{jr}$$

and

$$\frac{\partial U}{\partial p^{iq}} = v_{iq} = \frac{1}{2}\frac{\partial p^{jr}}{\partial p^{iq}} v_{jr} + \frac{1}{2}\frac{\partial v_{jr}}{\partial p^{iq}} p^{jr}$$

Since

$$\frac{\partial p^{jr}}{\partial p^{iq}} = 1 \quad \text{for } j, r = i, q \quad \text{or zero otherwise}$$

then

$$v_{iq} = \frac{1}{2} v_{iq} + \frac{1}{2} \frac{\partial v_{jr}}{\partial p^{iq}} p^{jr}$$

$$v_{iq} = \frac{\partial v_{jr}}{\partial p^{iq}} p^{jr}$$

$$v_{iq} = D_{ijqr} p^{jr}$$

where the quadri-valent holor D_{ijqr} is known as the "flexibility tensor". It possesses all properties of K_{ijqr} in inverse.

$$K^{misq} D_{mjsr} = \partial^{iq}_{jr}$$

where the bivalent version of δ is known as "the Kronecker delta".

Because of the dualities of Eqs. (11) and (11a), we shall omit the properties of D in these lectures. These two equations represent the two basic methods in discrete mechanics, i.e., the displacement (15) and the force methods (16). Consequently, the displacement (17) and the stress (18) functions play a very important role in these two equations. Depending upon the physical and geo, etrical characters of the problem, either one of these methods may possess some advantage over the other. Certain researchers (19), on the other hand, succesfully combine the two in one problem under the name of the "mixed method".

POSITIVE DEFINITIVENESS

The stiffness tensor, in addition to being a "Jacobian" as mentioned in (3), is positive and definite. For instance,

$$U = \tfrac{1}{2} p^i v_i \quad \text{and} \quad p^i = k^{ij} v_j$$

then

$$U = \tfrac{1}{2} k^{ij} v_i v_j$$

where v_i, v_j are arbitrary and $U>0$ then $k^{ij}>0$. Because of this property as well as the Jacobian properties of this entity, none of the "principal minors" of it vanish. This is very important when the final equation of large systems is solved by partitioning. Since

$$\left|k^{ii}\right|, \left|k^{jj}\right| > 0$$

there would be no need to test whether the sub-matrices on the main diagonal are singular or not. The unnecessary testings might be very costly.

ASSEMBLY OF K

When a continuous body is represented by interconnected elements, before introduction of boundary conditions, certain conditions must be satisfied between the elements. Such conditions which are referred to as "displacement compatibilities" often meet with lower degrees at the interfaces than at the nodal points. This is, of course, is the essence of the discrete body versus continuous. In conventional matrix methods, for instance, the assembly of elements is done as follows.

Let v_e and V respectively represent nodal point displacements of an element and the nodal point displacements of the entire body. Evidently there must be one to one correspondance between these two vectorial entities

(12)
$$v_e = AV$$

in which matrix A is referred to as "the connection matrix". Substituting this into Eq. (5) and introducing Castigliano's theorem, one obtains

$$U = \tfrac{1}{2} \int_\Omega V^* A^* b^* E b A V \, d\Omega$$

(13)
$$\frac{\partial U}{\partial V} = P = A \int_\Omega b^* E^* b \, d\Omega \, A \, V$$

where the integration is done over each element and

$$k_e = \int_\Omega b^* E b \, d\Omega$$

is referred to as "unassembled stiffness matrix" of the body. Finally, Eq. (13) indicates that the assembled stiffness of the body is

(14)
$$K = A^* k_e A$$

since it relates generalized vectors P and V representing nodal point forces and displacements respectively.

In spite of its neatness, the assembly of the stiffness matrix in this fashion lacks insight for the problem. Besides operating on larges matrices such as A and k_e any element of K is not explicitely expressed unless the triple multiplication in Eq. (14) is executed in proper order. Instead, we shall now assemble the

Assembly of Definitiveness

same equation in tensor notations without losing contact with the problem. Before doing that, it is necessary to remember that the "dyadic product" of univalent holors (vectors) results in a bivalent holor (matrix)[10].

$$u_i \, v_j = \omega_{ij}$$

If i and j take the value of nodal point designation, then, ω_{ij} would identify the proper location of element stiffness matrices within the assembled stiffness matrix. Then we shall see that Eq. (14) can be written as

$$K^{ijqr} = \gamma^{ij}_{ab} \, k^{abqr} \quad \begin{aligned} \gamma^{ij}_{ab} &= 1 \quad \text{for } i, j = a, b \\ &= 0 \quad \text{otherwise} \end{aligned} \qquad (15)$$

which indicates that the assembled stiffness matrix is the summation of the element stiffnesses.

In order to demonstrate the use of Eqs. (14) and (15), let us consider a small example shown in Fig. 1.

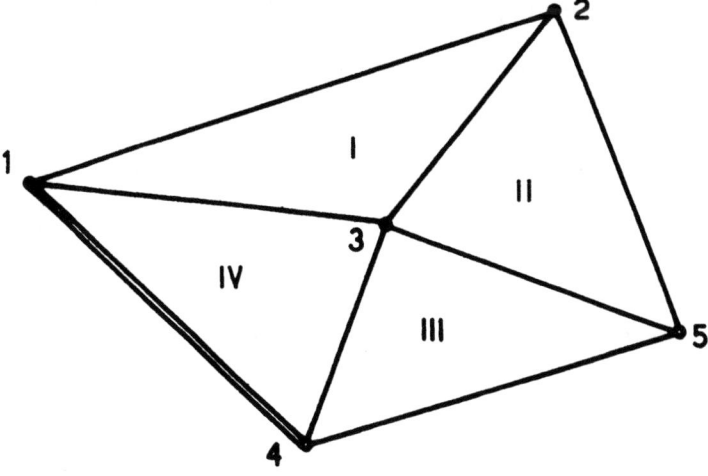

Fig. 1.

Eq. (14) for this example becomes

$$K = \begin{bmatrix} 1 & 0 & 0 & 0 & 0 & 0 & 0 & 0 & 0 & 1 \\ 0 & 0 & 1 & 0 & 0 & 1 & 0 & 0 & 0 & 0 \\ 0 & 1 & 0 & 1 & 0 & 0 & 0 & 0 & 1 & 0 & 0 \\ 0 & 0 & 0 & 0 & 0 & 0 & 1 & 0 & 0 & 1 & 0 \\ 0 & 0 & 0 & 0 & 1 & 0 & 0 & 1 & 0 & 0 & 0 \end{bmatrix} \begin{bmatrix} k_I & & & \\ & k_{II} & & \\ & & k_{III} & \\ & & & k_{IV} \end{bmatrix} \begin{bmatrix} 1 & 0 & 0 & 0 & 0 \\ 0 & 0 & 1 & 0 & 0 \\ 0 & 1 & 0 & 0 & 0 \\ 0 & 0 & 1 & 0 & 0 \\ 0 & 0 & 0 & 0 & 1 \\ 0 & 1 & 0 & 0 & 0 \\ 0 & 0 & 0 & 1 & 0 \\ 0 & 0 & 0 & 0 & 1 \\ 0 & 0 & 1 & 0 & 0 \\ 0 & 0 & 0 & 1 & 0 \\ 1 & 0 & 0 & 0 & 0 \end{bmatrix}$$

in which the diagonal matrix on the right is the <u>unassembled stiffness</u> matrix of the body.

If all the elements of this system and forces acting on it are in the same plane (plane stress problem) then, with the exception of k_{IV}, the element stiffness matrices are 6 by 6 (the explicite version of a triangular element stiffness matrix is presented later on). The order of K_{IV}, however, is 4 since there are only two nodal points with two degrees of freedom associated with this element.

In order to comply with Eq. (12), the element stiffness matrices must all be expressed in a common coordinate system otherwise the connection matrix will not consist of zeros and ones only. Such a coordinate system is often referred to as "the global coordinate system". It is, on the other hand, easier to develop the element stiffnesses in the special coordinate systems which might be oriented in a certain manner for each element.

Regardless of how they are oriented, evidently, they must be transformed into the global system before they are substituted into Eq. (14). Such a transformation can be best done in tensor notations. Let for instance x' and x respectively represent the local and global coordinate systems, then the element

Assembly of Definitiveness

stiffness matrices in global coordinates become

$$k^{ijqr} = k^{i'j'q'r'} \frac{\partial x_q}{\partial x_{q'}} \frac{\partial x_r}{\partial x_{r'}} \qquad (15a)$$

in which $\dfrac{\partial x_q}{\partial x_{q'}}$ and $\dfrac{\partial x_r}{\partial x_{r'}}$ are referred to as the rotation matrices. If both coordinate systems (local and global) are orthogonal, then

$$\frac{\partial x_q}{\partial x_{q'}} = \frac{\partial x_r}{\partial x_{r'}} = \frac{\partial x_{r'}}{\partial x_r}$$

which is known as "transpose = inverse" in matrix algebra. Because of such important properties of rotation matrices, very often the orthogonal coordinate axes are employed for both systems. This by no means eliminates the use of curvilinear or oblique coordinate systems. As a matter of fact, for certain problems curvilinear coordinates possess an advantage over the orthogonal systems. In such cases the tensorial transformation rules facilitate greatly the formulation of the problems.

In order to orient the local coordinate axes uniquely, the elements need to be identified by the nodal point numbers and their sequence. Assume, for instance, the elements are identified as

$$\begin{array}{rl} \text{I} & -\ 1\ 3\ 2 \\ \text{II} & -\ 3\ 5\ 2 \\ \text{III} & -\ 4\ 5\ 3 \\ \text{IV} & -\ \ \ 1\ 4 \end{array}$$

If a, b, c respectively represent the first, second and third number in the element designation, then, a typical stiffness matrix of a line and a triangular element can be partioned as

$$k = \begin{bmatrix} k^{aa} & k^{ab} \\ k^{ba} & k^{bb} \end{bmatrix} \qquad k = \begin{bmatrix} k^{aa} & k^{ab} & k^{ac} \\ k^{ba} & k^{bb} & k^{bc} \\ k^{ca} & k^{cb} & k^{cc} \end{bmatrix}$$

in which sub-matrices are of order 2.

Substituting these into Eq. (14) and executing the triple matrix product indicated above, the assembled stiffness matrix of a body in this example becomes

$$K = \begin{bmatrix} k_I^{aa} + k_{IV}^{bb} & & & & \\ k_I^{ca} & k_I^{cc} + k_{II}^{cc} & & \text{symm.} & \\ k_I^{ba} & k_I^{bc} + k_{II}^{ac} & k_I^{bb} + k_{II}^{aa} + k_{III}^{cc} & & \\ k_{IV}^{ab} & 0 & k_{III}^{ac} & k_{III}^{aa} + k_{IV}^{aa} & \\ 0 & k_{II}^{bc} & k_{II}^{ba} + k_{III}^{bc} & k_{III}^{ba} & k_{II}^{bb} + k_{III}^{bb} \end{bmatrix}$$

Examination of this matrix indicates certain facts. For example, the stiffness of joint 1 is made of stiffnesses of elements I and IV ; that of joint 3 is contributed by I, II and III; no member connects joints 1 and 5 to 4 etc.

In spite of being a common practice, the assembly of the stiffness matrix in this fashion (Eq.) (14) is not so efficient. It makes use of large matrices

Assembly of Definitiveness

(A and A* the connection matrices) which must partially or totally be stored in the machine. Instead, however, the assembled stiffness matrix K can be obtained from the individual element stiffnesses as indicated in Eq. (15). The identification of element stiffness matrices follows the rule of the dyadic product of vectors. According to the numbering sequence indicated above, the stiffness matrices of elements II and III, for instance, will be labeled as

$$k_{II} = \begin{bmatrix} k^{33} & k^{35} & k^{32} \\ \boxed{k^{53}} & k^{55} & k^{52} \\ k^{23} & k^{25} & k^{22} \end{bmatrix} \qquad k_{III} = \begin{bmatrix} k^{44} & k^{45} & k^{43} \\ k^{54} & k^{55} & \boxed{k^{53}} \\ k^{34} & k^{35} & k^{33} \end{bmatrix}$$

Consequently, K^{53}, for instance, is made of submatrices indicated in circles.

$$K^{53} = k^{53}_{II} + k^{52}_{III}$$

or

$$K^{53} = k^{ba}_{II} + k^{bc}_{III}$$

which was obtained previously by Eq. (14). In other words, the index manipulation in Eq. (15) takes care of the assembly of K without requiring any auxiliary storage in the machine. Neither unassembled stiffness matrix nor the connection matrices are needed. The element stiffness matrices are formed one at a time and then they are stored in K according to the indices with which they are associated.

Whether the elements are mixed i.e. triangular, rectangular, linear etc. or are in plane stress, in plate bending or in shell makes no difference. Their assembly is done without any additional precautions. For instance, the stiffness of nodal point 4 in this example is made of the stiffnesses of element III (a triangular element) and element 4 (a line element) as

$$K^{44\,qr} = k^{44\,qr}_{III} + k^{44\,qr}_{IV}$$

where

$$k_{III} = \begin{bmatrix} k^{44\,qr} & X & X \\ \hline X & X & X \\ \hline X & X & X \end{bmatrix} \qquad k_{IV} = \begin{bmatrix} X & X \\ \hline X & k^{44\,qr} \end{bmatrix}$$

Since the indices i and j assume value of 4 in elements III and IV only, $K^{44\,qr}$ is not effected by the stiffnesses of other elements, i.e., $\delta^{ij}_{ab} = 1$ for elements III and IV only, zero otherwise.

Equation (15) on the other hand indicates the same thing since 3 and 5 are the indices which occur in elements II and III only. Furthermore, Eq. (15) indicates that $K^{33\,qr}$ (the stiffness of nodal point 3) is not effected by the stiffness of element IV since i and j do not become 3 for element IV. Finally, through this equation, we are able to pre-judge the effects of problem-modifications on the assembled stiffness matrix. Such foresight plays an important role at the design stages.

RANK AND DEGENERACY OF STIFFNESS MATRIX.

In order to determine the degree of degeneracy and the rank of the stiffness matrices let us consider the equilibrium of the system. Since the loads acting on the system are in equilibrium, the following equation must always be satisfied.

$$H^r_{iq} P^{iq} = 0 \qquad (16)$$

in which H^r_{iq} (transfer holor), as we shall refer to it, represents linear transformation of forces from one point to another[11]. It becomes Kronocker delta in articulated systems in which forces do not possess angular components. For such systems, then, Eq. (16) represents simple additions.

In general, the transfer holor is in the following form :

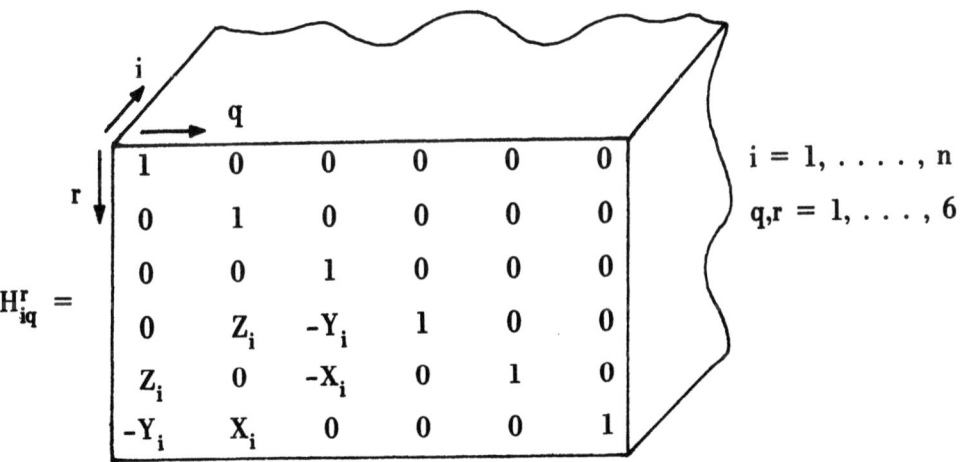

$$H^r_{iq} = \begin{bmatrix} 1 & 0 & 0 & 0 & 0 & 0 \\ 0 & 1 & 0 & 0 & 0 & 0 \\ 0 & 0 & 1 & 0 & 0 & 0 \\ 0 & Z_i & -Y_i & 1 & 0 & 0 \\ Z_i & 0 & -X_i & 0 & 1 & 0 \\ -Y_i & X_i & 0 & 0 & 0 & 1 \end{bmatrix} \quad \begin{array}{l} i = 1, \ldots, n \\ q,r = 1, \ldots, 6 \end{array}$$

in which X_i, Y_i, Z_i represent the coordinates of nodal point i.

By virtue of Eq. (11) the above equation (16) can be written as

$$H^r_{iq} P^{iq} = H^r_{iq} K^{ijqr} V_{jr} = 0$$

and knowing that V_{jr} can take any arbitrary value, we may conclude that

$$H^r_{iq} K^{ijqr} = 0 \qquad \begin{array}{l} i, j = 1, \ldots, n \\ q, r = 1, \ldots, N \end{array} \qquad (17)$$

This equation indicates that the total stiffness matrix of any structure prior to the introduction of boundary conditions is always "degenerate" and the "degree of degeneracy" is equal to q, r value, i.e., degree of freedom of a typical nodal point in general. Furthermore, this equation implies that the body is kinematically unstable, i.e., subject to a rigid body motion which becomes important for the analysis of "movable systems" such as ships, aircrafts etc...

Once the boundary conditions are introduced

$$V_{jr} = 0, c \qquad r \geqslant N$$

the remaining parameters become independent and the stiffness tensor shows all the characteristics of Jacobians.

Considering that the stiffness tensor is banded to begin with i.e.,

$$k^{ijqr} = 0 \qquad i, j > b$$

where b represents the "band width", Eq. (17) remains valid in certain regions even after the boundary conditions are introduced.

$$H^r_{iq} K^{ijqr} = 0 \qquad \begin{array}{l} i, j = s + a, \ldots, n \\ q, r = 1, \ldots, N \end{array} \qquad (18)$$

where s and a represent total number of prescribed boundary conditions and the nodal points adjacent to the boundaries. This equation indicates that the merates of the stiffness tensor in any row or column remain linearly dependent of each other.

Before going any further let us return to Eq. (17). This equation is valid for an element — as well as for the entire body — in which case i, j assume nodal designations and q, r vary according to degree of freedom of the nodal points. For a triangular element in plane stress problems, for instance, i, j = 1, 2, 3 and q, r = 1, 2. Therefore, element stiffness tensor in plane stress problem can at the most have ten independent parameters instead of twenty-one as often computed. Such an important reduction in the number of merates of an element stiffness tensor can be utilized in reducing the computation time and in increasing the accuracy of the results. For instance, the following bivalent version of the stiffness tensor of a triangular element in plane stress problem[19] illustrates explicitly the <u>ten</u> independent parameters and the validity of Eq. (17).

$$k = C \begin{bmatrix} y_{32}^2 & & & & & & \\ -\nu y_{32} x_{32} & x_{32}^2 & & & \text{symm} & & \\ \hline -y_{32} y_{31} & \nu x_{32} y_{31} & y_{31}^2 & & & & \\ \nu y_{32} x_{31} & -x_{31} x_{32} & -\nu y_{31} x_{31} & x_{31}^2 & & & \\ \hline y_{32} y_{21} & -\nu x_{32} y_{21} & -y_{31} y_{21} & \nu x_{31} y_{21} & y_{21}^2 & \\ -\nu y_{32} x_{21} & x_{32} x_{21} & \nu y_{31} x_{21} & -x_{31} x_{21} & -\nu y_{21} x_{21} & x_{21}^2 \end{bmatrix}$$

where

$$C = \frac{Et}{4A_{123}(1-\nu^2)}, \quad y_{ij} = y_i - y_j, \quad x_{ij} = x_i - x_j$$

Similarly, the bivalent version of the assembled stiffness tensor of the entire body before and after the boundary conditions are introduced (in Eq. (17) and Eq. (18) respectively) can be illustrated as

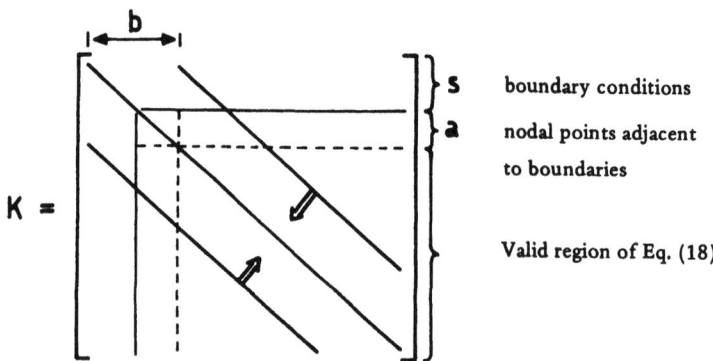

According to Eq. (18), then, the merates in any row or column where i, j > a + s are linearly dependent. Therefore the submatrices at the extreme diagonals of this region should vanish by simple additions of rows and columns. For problems in which the band width is three, there should be no need for inversion or elimination. The entire matrix in this region would be diagonalized by simple additions. The diagonalization of stiffness tensors with such predetermined operators shown in Eq. (18) is far from being finalized. This is hoped to be considered a single step, further steps yet to come. Eventually, uncoupling between the interior nodal points and those adjacent to the boundaries will be accomplished. Let us now consider the final equation in matrix notation

(19) $$P = K V$$

Multiplying both sides of this equation by H

$$HP = H K V$$

and post- and pre-multiplying K and V by H^* and H^{*-1} respectively results in

$$HP = H K H^* H^{*-1} V \tag{20}$$

or

$$P' = K'V' \tag{21}$$

where the band width in K′ is one (in sub-matrices'term) less than that of K. If the original band width was larger than 3, K′ of Eq. (21) is not uncoupled. Therefore, further relationships between the merates of K need to be determined. Each relationship reduces the band width by 2 (one from each side in sub-matrices term). Eventually the band width shrinks to 1 (diagonalization).

$$\underbrace{H_m \ldots H_3 H_2 H_1 P}_{\bar{P}} = \underbrace{H_m \ldots H_3 H_2 H_1 K H_1^* H_2^* \ldots H_m^*}_{\bar{K}} \underbrace{H_m^{*-1} \ldots H_2^{*-1} H_1^{*-1} V}_{\bar{V}} \tag{22}$$

where

$$\bar{K} = \begin{bmatrix} \begin{array}{cccc} x & x & . & x \\ x & x & . & x \\ . & . & . & . \\ x & x & . & x \end{array} & & \\ & \begin{array}{c} x \\ & x \\ & & x \\ & & & . \\ & & & & . \\ & & & & & x \end{array} & \end{bmatrix} \begin{array}{c} \Big\} a \\ \\ \Big\} n-(a+s) \end{array}$$

As a result of Eq. (22), the free joint displacements can be obtained without any inversion or elimination. Those adjoint to the boundaries, however, require inversion of a matrix of order a.

The solution of Eq. (22) results in $\bar{V}$ from which the unknowns of the original equation can be obtained as

(23) $$V = H_m^* \ldots H_2^* H_1^* \bar{V}$$

For the sake of simplicity, let us illustrate this phenomenon for a set which is tridiagonal in terms of 2 by 2 submatrices. Notice that the number of equations has no importance.

NUMERICAL EXAMPLE

Assume that the stiffness matrix equation of a system after the boundary conditions are introduced is in the following form. Determine the unknown nodal point displacements by Eq. (18).

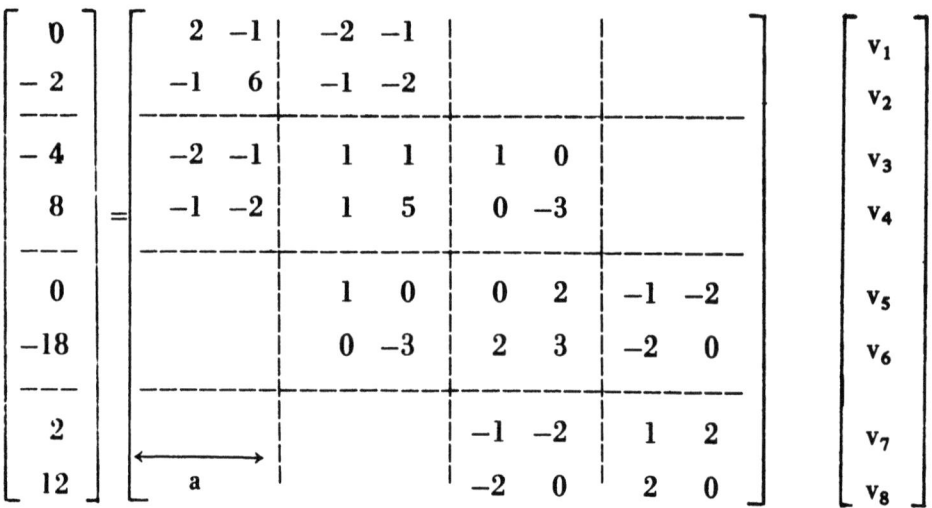

Step 1 : Add on every row (in submatrices term) all the rows that follow it. Do the same thing on the left-hand side.

Numerical Example

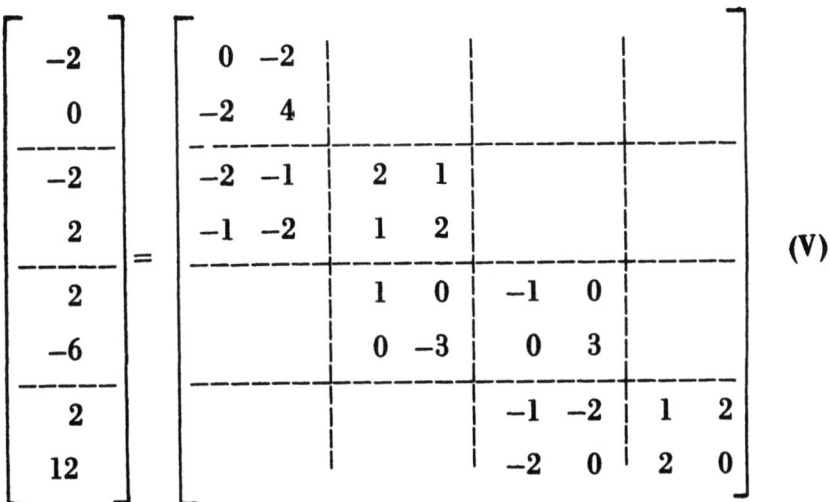

Step 2 : Add on every column all the columns that follow it. At the end of this step the stiffness matrix will be diagonalized (in submatrices term).

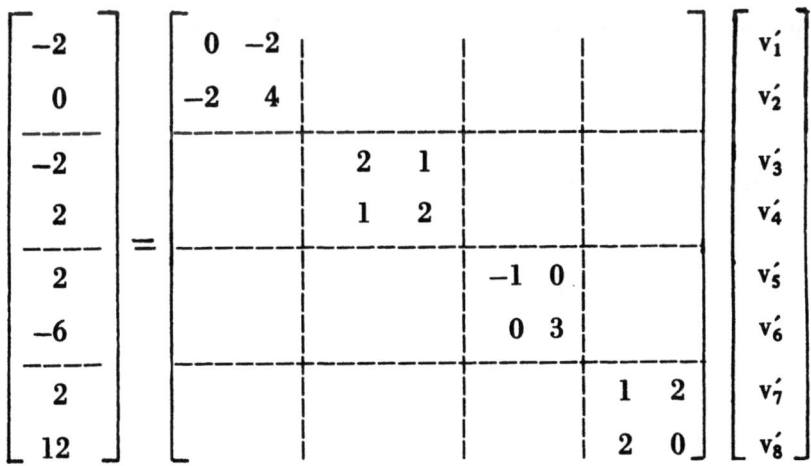

Step 3 : Solve the above equation by inverting submatrices on the main diagonal, then sum up the results downward, i.e., $v_3 = v_1' + v_2'$, etc.

$$\mathbf{v'} = \begin{bmatrix} 2 \\ 1 \\ \hline -2 \\ 2 \\ \hline -2 \\ -2 \\ \hline 6 \\ -2 \end{bmatrix}, \quad \mathbf{v} = \begin{bmatrix} 2 \\ 1 \\ \hline 0 \\ 3 \\ \hline -2 \\ 1 \\ \hline 4 \\ -1 \end{bmatrix}$$

where v´ is the solution vector for the set in step 2 and v is the answer to the problem.

This example indicates that in tri-diagonal systems the results can be obtained by simply adding rows and columns. The same procedure reduces the band width by one (in submatrices term) for other systems. Any classical method (elimination, inversion or iteration) for the solution of this kind of equations would require much more time and labor.

ILLUSTRATIVE EXAMPLE

The following example will illustrate the assembly of the stiffness matrix (Eq. 15) and the solution of the final equation (Eq. 19) for a structure consisting of line elements (the cantilever beam shown in Fig. 2). The purpose of chosing such a simple structure is to illustrate to the reader every detail of the procedure. For larger and complicated systems the procedure is the same but lengthier.

Illustrative Example

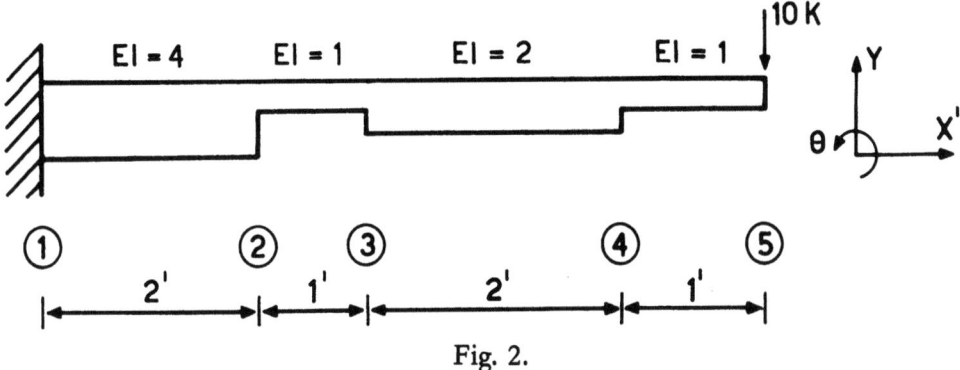

Fig. 2.

The complete stiffness matrix of a prismatic line element by neglecting the axial deformations is

$$k = \begin{bmatrix} k_{ii} & k_{ij} \\ \hline k_{ji} & k_{jj} \end{bmatrix} = \begin{bmatrix} a & & & \\ -b & c & \text{symm.} & \\ \hline -a & b & a & \\ -b & \frac{c}{2} & b & c \end{bmatrix} \qquad \therefore \quad \begin{aligned} a &= \frac{12EI}{L^3} \\ b &= \frac{6EI}{L^2} \\ c &= \frac{4EI}{L} \end{aligned}$$

which indicates that there are only 3 independent parameters (one half of the submatrices). This is due to the fact that a line element similar to the entire system is also subject to a rigid body motion.

$$h\,k = 0 \quad \therefore \quad h = \begin{bmatrix} 1 & 0 \\ -L & 1 \end{bmatrix}$$

Likewise,

$$\begin{bmatrix} 1 & 0 \\ 0 & 1 \end{bmatrix} \begin{bmatrix} a & -b \\ -b & c \end{bmatrix} + \begin{bmatrix} 1 & 0 \\ -L & 1 \end{bmatrix} \begin{bmatrix} -a & b \\ -c & \frac{c}{2} \end{bmatrix} = \begin{bmatrix} 0 & 0 \\ 0 & 0 \end{bmatrix}$$

If the segments are identified as 1-2, 2-3, etc., which is arbitrary, the assembly of the complete stiffness matrix can be done as explained previously.

The stiffness of nodal point 2, for instance, is

$$K^{33qr} = k_{II}^{33qr} + k_{III}^{33qr}$$

for which

$$k_{II} = \begin{bmatrix} k^{22} & k^{23} \\ k^{32} & \boxed{k^{33}} \end{bmatrix}, \quad k_{III} = \begin{bmatrix} k^{33} & k^{34} \\ k^{43} & k^{44} \end{bmatrix}$$

Therefore,

$$K^{33qr} = \begin{bmatrix} a & b \\ b & c \end{bmatrix}_{II} + \begin{bmatrix} a & -b \\ -b & c \end{bmatrix}_{III}$$

According to the given numerical data, the stiffness matrices of individual elements become

$$k_I = \begin{bmatrix} k^{11} & k^{12} \\ k^{21} & k^{22} \end{bmatrix} = \begin{bmatrix} 6 & & & \\ -6 & 8 & \text{symm.} & \\ -6 & 6 & 6 & \\ -6 & 4 & 6 & 8 \end{bmatrix}$$

$$k_{II} = \begin{bmatrix} k^{22} & k^{23} \\ k^{32} & k^{33} \end{bmatrix} = \begin{bmatrix} 12 & & & \\ -6 & 4 & \text{symm.} & \\ -12 & 6 & 12 & \\ -6 & 2 & \boxed{6} & 4 \end{bmatrix} = k_{IV}$$

$$k_{III} = \begin{bmatrix} k^{33} & k^{34} \\ k^{43} & k^{44} \end{bmatrix} = \begin{bmatrix} 3 & & & \\ -3 & 4 & & \\ \boxed{-3} & 3 & 3 & \\ -3 & 2 & 3 & 4 \end{bmatrix}$$

Illustrative Example

and the assembled stiffness matrix prior to the boundary conditions becomes

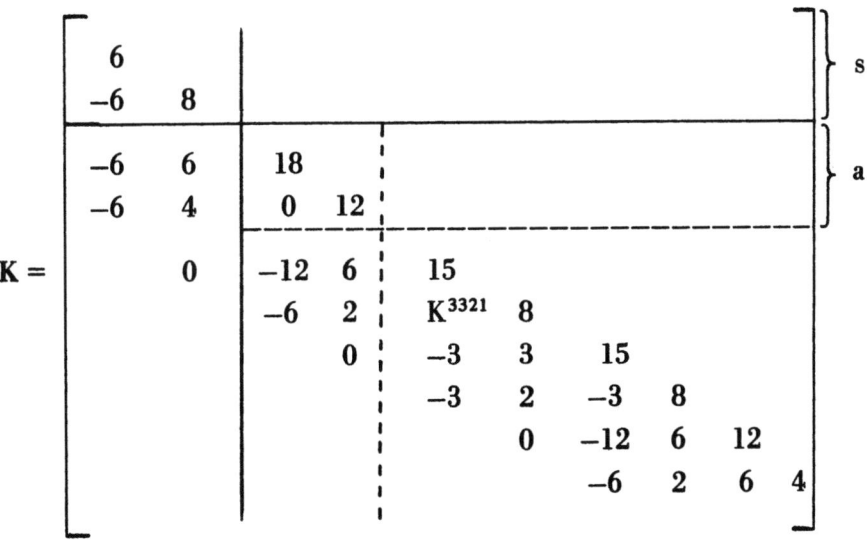

The stiffness matrix is assembled by the dyadic product of vectors identifying the elements. For instance,

$$K^{3321} = k^{3321}_{II} + k^{3321}_{III}$$
$$= 6 + (-3) = 3$$

The lower right corner of this matrix (in solid lines) represents the final stiffness matrix after the introduction of the boundary conditions. Notice that the boundary conditions erase only the first two rows (columns) of this entity since only one joint (nodal point 1) with two degrees of freedom is subject to an absolute restriction. Furthermore, $a = 2$ since there is only one joint (nodal point 2) adjacent to the natural boundaries. It is also worth to observe the validity of Eq. (17) for the entirety of this entity. Once the boundary conditions are introduced, however, such a relationship Eq. (17) is no longer valid in region a of this entity. The remaining portion (in dotted lines), on the other hand, still observes the same relationship. Consequently, Eq. (17) can properly be employed to diagonalyze this region of the entity. Although in this example the orders of this region and that of region a are 6 and 2 respectively, for most problems in practice the difference between the two is much larger. Therefore, the diagonalization (H matrices) deserves further attention.

The bivalent version of the transfer holor in this example is

$$H = \begin{bmatrix} 1 & 0 & 1 & 0 & 1 & 0 & 1 & 0 \\ -2 & 1 & -3 & 1 & -5 & 1 & -6 & 1 \\ & & 1 & 0 & 1 & 0 & 1 & 0 \\ & & -3 & 1 & -5 & 1 & -6 & 1 \\ & & & & 1 & 0 & 1 & 0 \\ & & & & -5 & 1 & -6 & 1 \\ & 0 & & & & & 1 & 0 \\ & & & & & & -6 & 1 \end{bmatrix}$$

If the final stiffness matrix is multiplied (pre- and post-) by H and H^* respectively as indicated in Eq. (20) the modified stiffness matrix K' and the modified load vector P' will become

$$K' = \begin{bmatrix} 6 & -6 & & & & & & \\ -6 & 8 & & & & & & \\ \hline & & 12 & -30 & & & & \\ & & -30 & 76 & & & & \\ \hline & & & & 3 & -12 & & \\ & & & & -12 & 49 & & \\ \hline & & & & & & 12 & 66 \\ & & & & & & -66 & 364 \end{bmatrix}, \quad P' = \begin{bmatrix} 10 \\ -60 \\ 10 \\ -60 \\ 10 \\ -60 \\ 10 \\ -60 \end{bmatrix}$$

Illustrative Example

The solution of this set yields

$$V' = \begin{bmatrix} -23.3 \\ -25 \\ -86.5 \\ -35 \\ -76.5 \\ -20 \\ -26.7 \\ -5 \end{bmatrix}$$

from which the final results can be obtained as (refer to Eq. 23)

$$V = H^* V' = \begin{bmatrix} 26.6 \\ -25 \\ 70 \\ -60 \\ 215 \\ -80 \\ 297 \\ -85 \end{bmatrix}$$

It can easily be verified that these results satisfy the original equation

$$P = K V = \begin{bmatrix} 0 \\ 0 \\ 0 \\ 0 \\ 0 \\ 0 \\ 10 \\ 0 \end{bmatrix}$$

which is the load vector on the system. The components of **V** represent the displacements and rotations of the free nodal points.

A classical solution of this problem, using conjugate beam method for instance agrees with the above results. Such a solution is illustrated in Fig. 3.

This example has demonstrated the assembly and the diagonalization of the stiffness matrix of a practical problem. Considering that the final stiffness matrix is of order 8, the solution of it is required inversion of matrices of order 2. H-operations are predetermined and need not be stored in the machine. The entire operation is uncoupled and can be done in parts. For instance, if the displacements of nodal point 5 (the end point) or any other point are desired, they can be obtained without solving the entirety of the problem.

$$\mathbf{K'} = \begin{bmatrix} 1 & 0 \\ -6 & 1 \end{bmatrix} \begin{bmatrix} 12 & 6 \\ 6 & 4 \end{bmatrix} \begin{bmatrix} 1 & -6 \\ 0 & 1 \end{bmatrix} = \begin{bmatrix} 12 & -66 \\ -66 & 364 \end{bmatrix}$$

$$\mathbf{P'} = \begin{bmatrix} 1 & 0 \\ -6 & 1 \end{bmatrix} \begin{bmatrix} 10 \\ 0 \end{bmatrix} = \begin{bmatrix} 10 \\ -60 \end{bmatrix}$$

$$\mathbf{V'} = \begin{bmatrix} -26.7 \\ -5 \end{bmatrix}$$

which represents the displacements of point 5 relative to point 4. At nodal point 2, on the other hand,

$$\mathbf{K'} = \left\{ \begin{bmatrix} 1 & 0 \\ -2 & 1 \end{bmatrix} \begin{bmatrix} 18 & 0 \\ 0 & 12 \end{bmatrix} + \begin{bmatrix} 1 & 0 \\ -3 & 1 \end{bmatrix} \begin{bmatrix} -12 & 6 \\ -6 & 2 \end{bmatrix} \right\} \begin{bmatrix} 1 & -2 \\ 0 & 1 \end{bmatrix} = \begin{bmatrix} 6 & -6 \\ -6 & 8 \end{bmatrix}$$

Illustrative Example

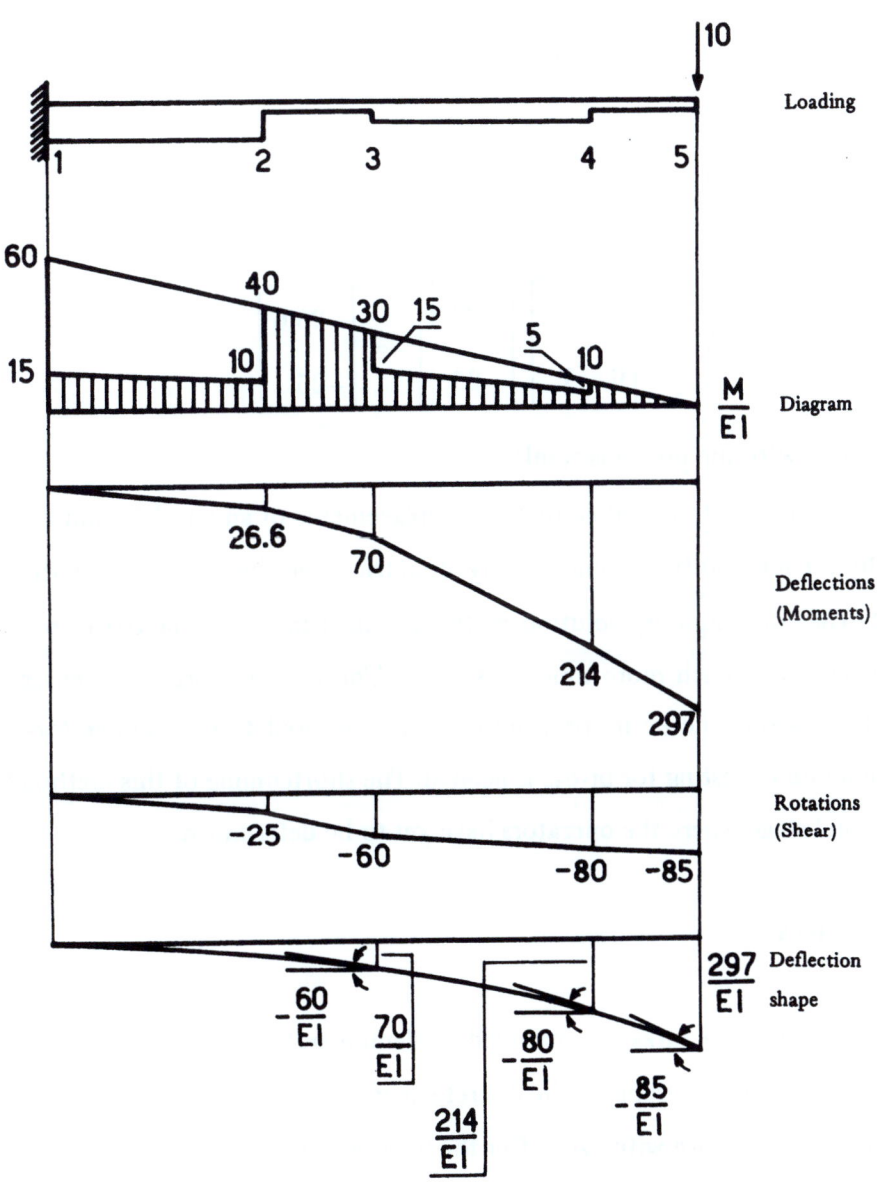

Fig. 3.

$$P' = \begin{bmatrix} 1 & 0 \\ -6 & 1 \end{bmatrix} \begin{bmatrix} 10 \\ 0 \end{bmatrix}$$

$$V' = \begin{bmatrix} -23.3 \\ -25 \end{bmatrix}$$

$$V = \begin{bmatrix} 1 & -2 \\ 0 & 1 \end{bmatrix} \begin{bmatrix} -23.3 \\ -25 \end{bmatrix} = \begin{bmatrix} 26.6 \\ -25 \end{bmatrix}$$

which agrees with the previous results.

It is worth to notice that the displacements of any nodal point are obtained by a triple multiplication of 2 by 2 matrices and inversion of a matrix of order 2. The contemporary solution of this problem involves a matrix of order 8 whose solution is much more time consuming. The most important advantage of this method is that the entire procedure is done by predetermined operators and no elimination or testing for pivots is needed. The shortcoming of this method lies in its second stage where the operators have yet to be determined.

CONCLUSION

These lectures are intended to demonstrate that the physical problems—— consequently problems in discrete mechanics—— can be formulated in tensor notations. As we witnessed after the middle of this century the yielding of analytical formulations to matrix manipulations, it is quite normal to expect that better formulations ought to come.

Since most of all physical entities are invariant under coordinate transformations and those in discrete mechanics are not any exception to this, their treatment as tensors instead of ordinary vectors and matrices deserves some justifications. Although every second order tensor can be classified and treated as a matrix, the reverse of course is not true. The beauty embedded in one can hardly be observed in the other. To take advantage of index manipulations and well defined coordinate transformations in tensors as opposed to the rigid rules and regulations of matrices may very well be the future trend of the finite element formulation of physical problems.

GLOSSARY

Holor : A Mathematical entity built up of several independent elements.

Tensor : A special type of holor which transform in one of the following ways.

Contravariant : $T^{i'j'k'\ldots\ell'} = \dfrac{\partial x^{i'}}{\partial x^i}\dfrac{\partial x^{j'}}{\partial x^j}\dfrac{\partial x^{k'}}{\partial x^k}\ldots\dfrac{\partial x^{\ell'}}{\partial x^\ell}\, T^{ijk\ldots\ell}$

Covariant : $T_{i'j'k'\ldots\ell'} = \dfrac{\partial x^i}{\partial x^{i'}}\dfrac{\partial x^j}{\partial x^{j'}}\dfrac{\partial x^k}{\partial x^{k'}}\ldots\dfrac{\partial x^\ell}{\partial x^{\ell'}}\, T_{ijk\ldots\ell}$

Mixed : $T^{i'j'k'\ldots\ell'}_{p'\ldots t'} = \dfrac{\partial x^{i'}}{\partial x^i}\dfrac{\partial x^{j'}}{\partial x^j}\ldots\dfrac{\partial x^p}{\partial x_{p'}}\ldots\dfrac{\partial x^t}{\partial x^{t'}}\, T^{ij\ldots\ell}_{p\ldots t}$

Merates : The elements of holor but do not necessarily imply the components of a vector.

Valence : The range of indices associated to a holor. Scalars, vectors, and matrices are respectively nilvalent, univalent and bivalent holors.

Indices: Are the superscripts and subscripts attached to a holor. A repeated index is said to be a dummy index.

Index Ballance: All the indices occur on each side of the equations must be ballanced both in letter and position.

Summation Convention: Whenever the same lateral index occurs as a superscript and as a subscript, i.e., a dummy index in a single term, summation is understood.

REFERENCES

1. Castigliano, A., Théorème de l'Equilibre des Systèmes Elastiques et ses Applications, Paris 1879.

2. Turnbull, H.W., The Theory of Determinants, Matrices and Invariants. Dover Publications, 1960.

3. Kardestuncer, H., Tensor properties of the Stiffness Matrix of Elastic Structures. IV IKM Congress in Weimar. DDR., 1967

4. Kardestuncer, H., Tensor in Discrete Mechanics. Tensor Quarterly, TSGB 1969 Great Britain.

5. Sokolnikoff, S., Mathematical Theory of Elasticity. John Wiley and Sons, New York 1946.

6. Brilloin, L., Les Tenseurs en Mécanique et en Elasticité. Masson et Cie, Paris 1936.

7. Borg, S.F., Matrix - Tensor Methods in Continuum Mechanics. Van Nostrand 1966.

8. Kardestuncer, H., Elementary Matrix Analysis of Structures, McGraw-Hill, in press.

9. Kron, G., Diakoptics. MacDonald, London, 1963.

10. Moon, Parry : Unpublished notes on Tensors.

11. Kardestuncer, H., Piping Flexibility Analysis by Matrices. Report No.CE66-3 University of Connecticut, 1966.

12. Kardestuncer, H., Analyse Matricielle des Structures et Application aux Ordinateurs. CNRS Publication, Paris 1963.

13. Moon and Spencer, Vectors. Van Nostrand 1965.

14 Rubinstein, M.F., Matrix Computer Analysis of Structures. Prentice-Hall 1966.

15 Livesley, R.K., Matrix Methods of Structural Analysis, Pergamon Press, Oxford, 1964.

16 Argyris, J.H., Energy Theorems and Structural Analysis, Butterworths Scientific Publications, London, 1955.

17 Turner, M.J. et al., Stiffness and Deflection Analysis of Complex Structures, Journal of Aero. Sci. 23, No. 9, 1956.

18 Pian, T.H.H., Derivation of Element Stiffness Matrices by Assumed Stress Distribution, AIAA J. 2, No. 7, 1964.

19 Gallagher, R.H. et al., Recent Advances in Matrix Structural Analysis and Design, University of Alabama Press, 1971.

20 Kron, Gabriel, Tensors for Circuits, Dover Publications, 1959.

21 Nadeau, Gerard, Introduction to Elasticity, Holt, Rinehart and Wiston, Inc., 1964.

22 Fung, Y.C., Foundations of Solid Mechanics, Prentice-Hall, New York, 1965.

23 Eringen, A.C., Mechanics of Continua, John Wiley & Sons, Inc., New York, 1967.

24 Sedov, L.I., Introduction to the Mechanics of Continuous Medium, Addison-Wesley, Boston, 1965.

25 Bor, S.F., The Tensors in Structural Theory, Annals of Academy of Science, New York, 1962.

26 Soule, W., Tensor Flexibility Closed-Loop Piping Systems, Journal of Applied Mechanics, March, 1958.

CONTENTS

Preface..	3
Introduction..	5
Matrix Formulation of the Problem...	8
Tensorial Formulation of the Problem...	9
Positive Definitiveness..	15
Assembly of K...	15
Rank and Degeneracy of Stiffness Matrix...	22
Numerical Example..	28
Illustrative Example...	30
Conclusion...	38
Glossary...	40
References...	41

If you have any concerns about our products,
you can contact us on
ProductSafety@springernature.com

In case Publisher is established outside the EU,
the EU authorized representative is:
**Springer Nature Customer Service Center GmbH
Europaplatz 3, 69115 Heidelberg, Germany**

Printed by Libri Plureos GmbH
in Hamburg, Germany